Adventure at Kitty Hawk

by Mark Fara

Requests for permission to make copies of any part of the work should be addressed to School Permissions and Copyrights, Harcourt, Inc., 6277 Sea Harbor Drive, Orlando, Florida 32887-6777. Fax: 407-345-2418.

HARCOURT and the Harcourt Logo are trademarks of Harcourt, Inc., registered in the United States of America and/or other jurisdictions.

Printed in Mexico

ISBN-13: 978-0-15-362487-2

ISBN-10: 0-15-362487-6

2 3 4 5 6 7 8 9 10   126   10 09 08

Harcourt
SCHOOL PUBLISHERS

Visit *The Learning Site!*
www.harcourtschool.com

Introduction

The world got a lot smaller in 1900. It did not actually shrink, but long distances were on their way to seeming shorter. That year, brothers Orville and Wilbur Wright built their first glider. A glider is a machine that flies on air currents. It looks like an airplane, except that it is smaller, lighter, and does not have an engine.

In 1902, the Wright brothers made improvements to their first glider. Then in 1903, they built one with an engine—the first true airplane. The first successful airplane flight was on December 17, 1903. For the next two years, they worked to improve the design. By 1908, the Wright brothers were making one- and two-hour flights in the United States and Europe.

Today, traveling by airplane seems routine to many people. Even long flights, like those from the United States to Australia, occur daily. But we should never forget the two men who started it all.

Today the Wright brothers' first airplane is displayed at the National Air and Space Museum at the Smithsonian Institution in Washington, D.C. A plaque below the airplane reads:

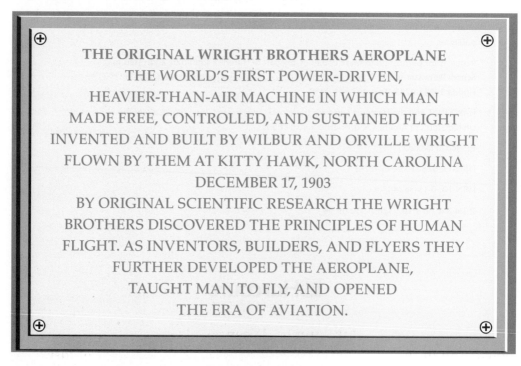

THE ORIGINAL WRIGHT BROTHERS AEROPLANE
THE WORLD'S FIRST POWER-DRIVEN,
HEAVIER-THAN-AIR MACHINE IN WHICH MAN
MADE FREE, CONTROLLED, AND SUSTAINED FLIGHT
INVENTED AND BUILT BY WILBUR AND ORVILLE WRIGHT
FLOWN BY THEM AT KITTY HAWK, NORTH CAROLINA
DECEMBER 17, 1903
BY ORIGINAL SCIENTIFIC RESEARCH THE WRIGHT
BROTHERS DISCOVERED THE PRINCIPLES OF HUMAN
FLIGHT. AS INVENTORS, BUILDERS, AND FLYERS THEY
FURTHER DEVELOPED THE AEROPLANE,
TAUGHT MAN TO FLY, AND OPENED
THE ERA OF AVIATION.

The First Airplane
Orville and Wilbur Wright made two flights each on
December 17, 1903. The longest flight lasted 59 seconds.

The Era of Aviation

Aviation is the science of heavier-than-air flight. Heavier-than-air refers
to a group of aircraft that includes gliders and airplanes. The plaque at
the Smithsonian Institution reminds us that the Era of Aviation started
with the Wright brothers. It is true that people had found ways to leave the
ground before the Wrights came along. It is also true that some people used
gliders before they did. But nobody had the kind of success that the Wright
brothers had. They took what people had done earlier, added their own
hard work and ideas, and changed the world.

The Problem with Flying: Forces

Gravity is the force that keeps everything on the ground. If there was
suddenly no gravity, everything on Earth—including the atmosphere—
would float off into space. When you put your backpack on the floor, gravity
keeps it there. When you put it on a hook, gravity makes it hang.

Gravity also causes things to fall. When you stumble on the sidewalk,
gravity pulls you toward the ground. When you drop a ball, gravity pulls it
down. And the longer something falls, the faster it goes and the harder
it lands.

There is no such thing as an "off" switch for gravity. So if you want to fly, you have to find a way to balance gravity's pull. You also need to figure out how to manage with gravity well enough to come back down safely.

Another name for the pull of gravity on a glider is *weight*. In order for a glider to lift up off the ground, something must apply a force upward greater than the glider's weight. This force is called *lift*. Think of a paper airplane. You supply the force, or *thrust*, to make it move forward. This movement also produces lift. Another force called *drag* acts against the plane, slowing it down. Drag is caused by the air interacting with the plane. As the plane slows, it loses lift and eventually falls.

Early Flight

Being able to fly has been a dream of humans, even in early civilizations. Have you ever flown a kite? The ancient Egyptians flew them, too—about 2500 years ago. In the 1100s the Chinese invented parachutes, which use air to help people fall slowly and safely.

About 500 years ago, an inventor named Leonardo da Vinci drew sketches of flying machines that look much like our airplanes today do. He never built one, but he did invent the propeller. The propeller would later make airplane flight possible.

Forces on a Plane

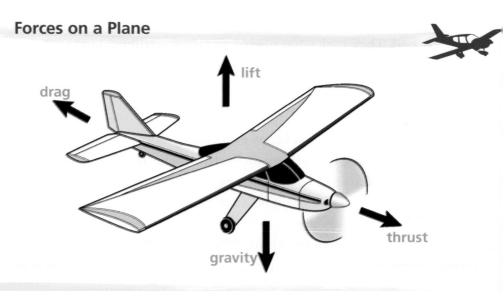

lift

drag

thrust

gravity

The first flights were made in lighter-than-air craft, so called because they use hot air or gases to become "lighter than air." In 1782, two brothers in France invented the first successful hot-air balloon. Balloons do not use engines to fly. In fact, balloons do not really fly at all, they float. A balloon is simply a huge bag with a basket hanging below it. The air inside the bag is heated. Because hot air is less dense than cooler air, the bag rises, taking the basket with it. Early balloonists often took a device called a brazier with them. A brazier is a metal container in which coal is burned to make heat and keep the air in the bag hot. In later years, some types of gas were discovered that are lighter than hot air. Helium and hydrogen are the two most common gases used in balloons. Helium is usually a better choice because hydrogen is flammable. Helium is the gas used in birthday balloons.

Balloons were used in both World War I and World War II. Today, they are used mostly for scientific research and for sport. Many places have balloon clubs, where members meet to ride in their balloons and have races. Some people also try to set records with their balloons. In 1999, two Europeans became the first people to go completely around the world in a balloon without landing. In 2002, an American named Steve Fossett became the first person to do it alone. It took him just under 15 days.

Zeppelin

An airship is another type of lighter-than-air craft. It is also sometimes called a blimp or a zeppelin. It is similar to a balloon but is usually much larger and has an oblong shape. Airships are usually filled with hydrogen or helium. They have large boxes called gondolas hanging beneath them to hold people. They use motors and propellers to move forward and backward but are not considered heavier-than-air because gas keeps them afloat. They were used in World War II, along with balloons. These days they are used mostly for advertising at outdoor events.

Thank You, Sir George

In the late 1700s, an Englishman named Sir George Cayley began studying the way that birds fly. For more than 50 years he tried to design and build machines that could do what birds do. In 1853, he built a big glider and put his assistant on it—and it flew! It went through the air for 900 feet before it crashed. This was the first time in recorded history that a human being flew any real distance in something heavier than air. (It was the first crash, too.) Before he died in 1857, Cayley wrote that long flights would not be possible until somebody invented the right kind of engine to power an airplane. His prediction turned out to be right.

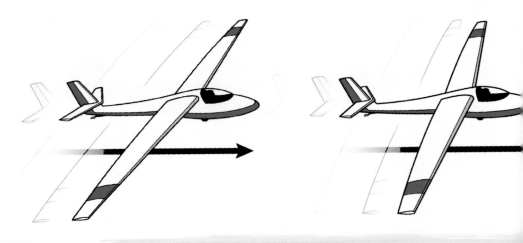

A glider has no motor. It flies by using wind.

Heavier Than Air?

For many years people thought that if we could only imitate the movements of birds, human flight would be possible. In a way, that is what the Wright brothers did.

Airplanes need their wings to fly, just as birds do. Some type of wing is needed for any object to fly. Wings slice through the air. When the wing slices at the proper angle, the air above and beneath the wing works to lift it higher. It uses the air moving over the surface of the wings to overcome gravity.

But flying requires more than wings. Have you ever watched a bird flap its wings and then simply hold them out straight? It continues to fly (similarly to the way a glider does) for a little while. Then it has to flap its wings again so that gravity does not pull it down.

Birds use their wings to push the air behind them, and this makes them go forward. The bird is creating thrust—the force that pushes an object through the air. Think about swimming. When you swim, you move your arms and kick your feet in a way that pushes the water behind you and makes you go forward. Birds do the same thing when they fly.

Now think of a paper airplane. When you throw it, your arm is giving it thrust. It has no thrust while it is flying, and that is why it does not go very far. If a paper plane could flap its wings like a bird, it would keep flying.

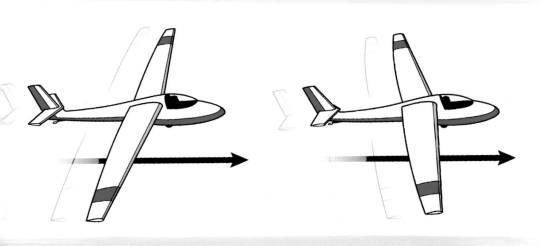

Airplanes do not flap their wings either. But because of the thrust they get from propellers or jet engines, they stay in the air. A propeller spins very rapidly, forcing air behind it and moving the plane forward. Jet engines create propulsion, forcing air outward to push the plane forward. To understand how a jet engine works, think of blowing up a balloon and letting it go.

Without air, wings are useless. A plane or a bird could not fly in outer space because there is no air to hold it up.

An Early Victory

Most educated people in the 1870s accepted the idea of gliders, but thought of the notion of an airplane as far-fetched. People who studied flight and tried to fly were considered by most to be dreamers. But then a man in Germany named Otto Lilienthal began studying the work of Sir George Cayley. In 1891, he experimented with his own gliders. Over the next five years, he made over 2000 flights. He improved on earlier designs and had some successes. People started paying attention. News of Lilienthal's work spread across the ocean and captured the interest of Orville and Wilbur Wright.

The Wright Brothers

Orville and Wilbur Wright

Orville and Wilbur Wright grew up in a family of five children. Wilbur was the older of the two, born on a farm in Indiana in 1867. In 1871, Orville was born in Dayton, Ohio. Their father was a bishop in the United Brethren Church. He changed jobs often, so the family moved frequently.

The brothers grew up in a strict but loving household. As an adult, Orville remembered his childhood fondly. He wrote: "We were lucky enough to grow up in an environment where there was always much encouragement to children . . . to investigate whatever aroused curiosity."

Although both brothers attended high school in Dayton, neither actually graduated. They were both very interested in mechanical things and taught themselves mathematics.

When Wilbur was 19 years old, someone accidentally hit him in the face with a bat while he was playing on the ice in wintertime. At first the injury did not seem serious, but he started having heart problems a few weeks later. He stayed home for the next four years because of his health and to care for his mother, who was dying of a lung disease. In 1892, the brothers started the Wright Cycle Company. There they designed, built, and sold bicycles for the next 10 years. They also studied aviation and engineering on their own.

By this time, there were many people in the United States and Europe who were trying to invent an airplane. Otto Lilienthal was killed in 1896 while he was testing a glider. A gust of wind damaged one of the glider's wings, and he fell about 60 feet. He died the next day. His last words were, "Sacrifices must be made."

Lilienthal's death taught the brothers an important lesson: It would be important to perfect the design of gliders before inventing an airplane. They began their work by learning as much as they could about what others had done. They studied many glider designs and determined what was best about each of them. They studied Lilienthal's work and used his insights in their own designs. Many other inventors who were trying to invent an airplane did not do their homework first. This is a large part of the reason that they failed and that the Wright brothers succeeded.

In 1899, the Wright brothers flew a kite equipped with special controls that could be used to steer it. This test gave them valuable information that they would use when it was time to build their airplanes.

The brothers tested their first glider in 1900. It was similar to the kite they had tested, only it was larger and had a seat for a pilot. They picked Kitty Hawk, North Carolina, for their test because it had the steady winds they needed. It also had large sand dunes that would provide a soft landing. Wilbur actually achieved a few seconds of controlled flight, but the test did not go as well as they had hoped. Still, it was one of the best gliders that had been built by anyone so far. The next year they returned to Kitty Hawk with a new glider. This one was longer and wider and had wings shaped differently from the first one. That test went well, but there was still much work to do.

Octave Chanute

A man from France named Octave Chanute had become a well-known and respected engineer by building bridges for railroads. In fact, he built the first bridge over the Missouri River. He retired from engineering in 1889 and devoted his time to the science of aviation. He was president of a respected engineers' club and experimented with his own gliders on the sand dunes in Indiana. He and his assistants used five different gliders and made about 1000 flights. He published the results in a series of articles in a railroad magazine. In 1894, all of his articles were published in a single book. At that time, it was one of the best books on flying that had been written. The Wright brothers were very familiar with Chanute's work and studied his ideas closely.

The Wright brothers and Chanute first met in 1900, the same year the brothers tested their first glider. Chanute encouraged them to continue their work because he was convinced that they were on the right track. In 1901, he visited them in Dayton and in Kitty Hawk.

Chanute encouraged the Wright brothers to publish the results of their experiments, and he invited Wilbur to talk about flying at Chanute's engineers' club. In his introduction of Wilbur, Chanute praised the Wright brothers, saying, "These gentlemen have been bold enough to attempt some things which neither Lilienthal . . . nor myself dared to do . . . and they have accomplished very remarkable results."

He believed strongly in sharing what he knew about flying, and he expected others to feel the same way. The Wright brothers wanted to protect their discoveries by having them patented. This difference led to some hard feelings between the Wrights and Chanute, but they worked out their disagreements. Wilbur gave the eulogy at Chanute's funeral in 1910.

The Work Continues

The brothers knew that it was too expensive to keep building gliders and taking them all the way to Kitty Hawk to test them. During the winter of 1901, they built a wind tunnel, a large tube through which air is blown. It duplicates the effects of flying without having to actually fly. The Wrights tested many different wing shapes in this tunnel and figured out which of them would work best on a real airplane.

In 1902, they took what they had learned from their wind tunnel tests and built their best glider to date. When they tested it, it flew 662 feet. They were so pleased that they decided it was time to take the next step.

They returned home and tested propellers, just as they had tested the wings. They searched for an engine that was big enough to power their airplane but light enough not to weigh it down. When they realized that nobody had invented an engine like that yet, they went ahead and built it themselves. During the later months of 1903, they decided that it was time for another trip to Kitty Hawk.

They knew that the test would go well, and it did. After a few false starts, Orville made a 12-second flight. Orville and Wilbur Wright had built the first successful airplane.

They continued taking turns on the new invention. They made a total of four flights that day. The longest one was made by Wilbur, and it lasted 59 seconds.

Success and Tragedy

There was more to come, of course. The brothers continued to test and improve their designs. In 1904, they made a total of 105 flights in their hometown of Dayton. By then, they were able to stay in the air for more than five minutes at a time. By 1905, they were staying in the air for 30 minutes or more. They also made improvements in controlling their airplanes and could steer and turn as they wished.

By now, others were starting to copy the brothers' invention. A man named Glenn Curtiss won prizes by flying planes made by a group in the United States. He also designed some of the earliest seaplanes. A company in France made planes and did experiments to improve them, as did companies in England and Germany.

Wilbur went to France and demonstrated one of his planes by making a well-controlled flight lasting more than two hours. At the same time, Orville demonstrated the brothers' invention for the United States Army. He took the world's first airplane passenger on a six-minute flight. A later flight with a passenger ended in tragedy when the plane crashed. Orville was injured, and the soldier with him died a few hours later. Despite the misfortune, testing continued, and the Army bought an airplane from the Wrights in 1909. It was the first successful military plane. Orders for more planes followed. Countries in Europe also wanted some for their armies. The brothers needed a place to manufacture their airplanes, so they started the Wright Company.

Wilbur died of typhoid fever in 1912. In his diary that day, Bishop Wright wrote, "This morning at 3:15, Wilbur passed away, aged 45 years, 1 month, and 14 days. A short life, full of consequences. An unfailing intellect, imperturbable temper, great self-reliance and as great modesty, seeing the right clearly, pursuing it steadily, he lived and died."

Fame and Misfortune

The Wrights became famous. After Wilbur died, Orville received many awards for the work he and his brother had done. There were problems, though. In 1906, the Wright brothers had patented the steering system they invented. A patent is a legal document that gives an inventor the right to be the only one to make money from an invention. Patents are only legal in the country that issues them. France and Germany refused to give the Wrights patents, so anyone in those countries could copy the brothers' inventions without paying for them.

Even in the United States, where they had a patent, problems arose. Glenn Curtiss had tried to join forces with the Wright brothers in 1906. When they declined, Curtiss teamed up with Alexander Graham Bell, the inventor of the telephone. The two of them formed a group that built airplanes. Curtiss had been using the Wrights' steering system in his airplanes without paying the Wrights anything for it. The Wrights sued him, and after Wilbur's death Orville received a large sum of money. The British government awarded some money too, from planes that had been made there. But none of these payments came close to what the Wrights were entitled to receive.

early biplane   water plane   fi

Orville sold the Wright Company in 1915, but his passion for flying lasted until his death in 1948. He continued research and worked for the U.S. government's National Advisory Committee on Aeronautics. In 1958, this committee became NASA, the government's space division.

What the Wrights Started

It is no exaggeration to say that the Wright brothers changed the world. People have built on what the Wrights invented, much like the brothers themselves did. That process continues to this day.

- In 1911, a man in a Wright plane made the first flight across the United States. He made several stops along the way and took a total of 84 days to finish his trip.
- World War I lasted from 1914 to 1918. For the first time, airplanes were used in a war. At first, each side used airplanes to spy on the other side. As the war continued, both sides used airplanes that carried machine guns and bombs to attack their enemies on the ground and in the air. Seaplanes were used to help defeat submarines.
- In 1919, two men from Britain became the first people to fly across the Atlantic Ocean without stopping. They made the trip in a little more than 16 hours. A newspaper in London awarded them $50,000 for their achievement.
- In 1925, Congress gave the Post Office Department permission to use airplanes to move mail. Over the next couple of years, mail was flown between major cities in the United States as well as into Canada and South America.

Stealth bomber   **modern airliner**

- Not until 1927 did someone fly solo across the Atlantic Ocean nonstop. Charles Lindbergh, an American, took between 33 and 34 hours to fly from New York City to Paris, France—a distance of more than 3500 miles!
- Between 1930 and 1940, the airplane industry began to emerge. Airplanes owned by private companies moved people and packages across the United States and over the oceans. By the start of World War II, almost 50 countries had regular airline service.
- In 1932, Amelia Earhart became the first woman to fly alone across the Atlantic Ocean. She set a new record for the trip—$13\frac{1}{2}$ hours. Both the United States and France honored her for this achievement. She went on to become the first woman to fly across the Pacific Ocean and to set a speed record from Mexico City to New York. In 1937, she and her navigator tried to fly completely around the world. They disappeared partway into their flight. The United States Navy tried to find them, but they were never heard from again.
- In 1941, the United States entered World War II. The United States quickly developed better airplanes to use in battle. The Germans began using jet-propelled combat airplanes in 1944. Toward the end of the war in 1944, almost 100,000 planes a year were being manufactured in the United States, mostly for military use.
- After the war, airplane improvements developed by the military started to be used in civilian planes. Planes became bigger and faster and could fly higher.

- The United States government launched *Columbia,* the first space shuttle, in 1981. There is no other craft quite like a shuttle. It flies like a spaceship when it is in space. When it returns to Earth it acts like a glider. It lands on a regular airstrip without the power of an engine of any kind.

Flight Today

Today there are jets that can travel faster and farther than the Wright brothers ever dreamed. We have airplanes that can take off from and land on ships. We have airplanes that operate on water, like flying boats. And who knows what aviation engineers will invent next? Whatever it is, we should never forget whom to thank for it. We will always be in debt to Orville and Wilbur Wright and their determination to succeed.

Space Shuttle